The Wizard's Workshop
A Science Activity Book

For Michael, Sarah, and Julia.
(The first witches and wizard to join my potions class.)
—Jennifer

Dedicated to my family. You guys are magical.
—Katie

Text © 2018 Jennifer K. Clark
Illustrations © 2018 Katie Payne
All rights reserved.

No part of this book may be reproduced in any form whatsoever, whether by graphic, visual, electronic, film, microfilm, tape recording, or any other means, without prior written permission of the publisher, except in the case of brief passages embodied in critical reviews and articles.

The opinions and views expressed herein belong solely to the author and do not necessarily represent the opinions or views of Cedar Fort, Inc. Permission for the use of sources, graphics, and photos is also solely the responsibility of the author.

ISBN 13: 978-1-4621-2167-0

Published by Plain Sight Publishing, an imprint of Cedar Fort, Inc.
2373 W. 700 S., Springville, UT, 84663
Distributed by Cedar Fort, Inc., www.cedarfort.com

LIBRARY OF CONGRESS CATALOGING-IN-PUBLICATION DATA

Names: Bones, Creek E., 1974- author. | Clark, Jennifer K., author.
Title: The wizard's workshop / Master Wizard Creek E. Bones ; assisted by
   Wizard in Training Jennifer K. Clark.
Description: Springville, Utah : Plain Sight Publishing, an imprint of Cedar
   Fort, Inc., [2018] | Includes bibliographical references and index.
Identifiers: LCCN 2017059712 | ISBN 9781462121670 (perfect bound : alk. paper)
Subjects: LCSH: Science--Experiments--Juvenile literature. |
   Liquids--Experiments--Juvenile literature. | Wizards--Juvenile literature.
Classification: LCC Q163 .B675 2018 | DDC 507.8--dc23
LC record available at https://lccn.loc.gov/2017059712

Cover and page design by Katie Payne and Shawnda T. Craig
Cover design © 2018 Cedar Fort, Inc.
Edited by Kaitlin Barwick

Printed in the United States of America

10 9 8 7 6 5 4 3 2 1

Printed on acid-free paper

The Wizard's Workshop
A Science Activity Book

Plain Sight Publishing
An imprint of Cedar Fort, Inc.
Springville, Utah

By the way, if you don't like me writing in this book, then don't read it!

Written by Master Wizard Creek E. Bones
and Apprentice Jennifer K. Clark
Illustrated by Katie Payne

This is a pass-along book.
When you find that you no longer need it, pass it along to another aspiring wizard.

Previous Owners:

Professor Humbleporf
Dan D. Lions
Ivonna Suk Yerblood ← *I think this person is a vampire!*
Luna Tick
Gene E. Yuss
Larry Blotter

This is me! And no, I am not that kid with the scar. (→ Larry Blotter)

Put your name here.

A Note to Parents *Boring stuff!*

THIS BOOK IS INTENDED TO encourage imaginative play. Science experiments are meant to be fun. Who wouldn't be interested in knowing that combining sodium bicarbonate and vinegar will produce carbon dioxide? $H_2CO_3 \longrightarrow H_2O + CO_2 \ldots$

Excuse me . . . wake up; I was talking here.

Imagine if you and your children could combine Dragon Spit with Powdered Unicorn Horn to create a potion that will protect your home from dragons. Suddenly you have a day filled with fantastical creatures and amazing magic!

The experiments in this book can take as little as 10 minutes but can easily be turned into hours of playtime.

So how do you turn your kitchen into the ultimate potions making classroom?

1. ENCOURAGE KIDS to dress the part. Break out the magic wand and cape.

2. GO TO the website (www.katiepayne.com/wizardsworkshop) and print the provided labels to turn ordinary kitchen ingredients into a magical substances. Dragon Spit is much more interesting than vinegar. Labels may be attached to original containers or you can make your own potion bottles by recycling old and unique containers.

3. PLAY ALONG. Half of the fun is adding your own imagination to the experiments.

Potion Labels

Go to www.katiepayne.com/wizardsworkshop

Understanding the Experiment

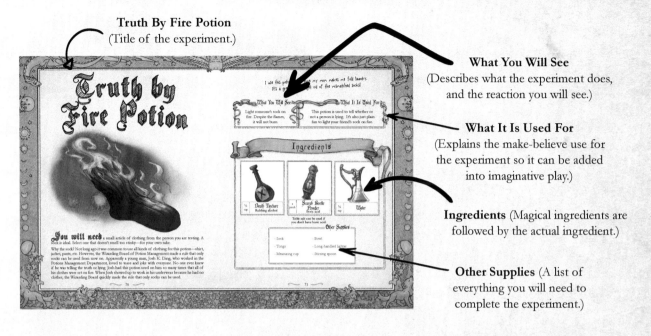

Truth By Fire Potion
(Title of the experiment.)

What You Will See
(Describes what the experiment does, and the reaction you will see.)

What It Is Used For
(Explains the make-believe use for the experiment so it can be added into imaginative play.)

Ingredients (Magical ingredients are followed by the actual ingredient.)

Other Supplies (A list of everything you will need to complete the experiment.)

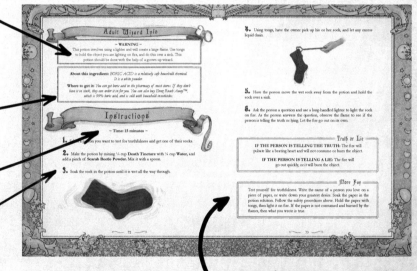

Adult Wizard Info (Anything in an experiment that should be dealt with carefully will be listed here.)

About These Ingredients (Explains what the unusual ingredient is and where it can be found.)

Time (This is the average time it takes to put the ingredients together. This does not include added make-believe time.)

Instructions (Step by step instructions for the experiment.)

More Fun/Experiments (If there are additional experiments or other fun things to do with an experiment, they will be listed here.)

About Ingredients

The experiments in this book were selected because of the magical reaction they produce. Sometimes a special ingredient is needed to provide the huge "wow" factor. While most ingredients can be easily found at local stores or around your house, harder-to-find ingredients will have suggestions of where you can get them. There is also a complete ingredients list at the end of the book along with where to purchase them.

V

Rules of Potion Making

1. A WIZARD IS ALWAYS SAFE *You don't win awards for blowing yourself up.*

Potion making can be hazardous. Adult supervision is advised at all times. Heed all warnings when making any potion, and use proper safety equipment such as eye protection and gloves whenever advised.

2. A WIZARD IS ALWAYS PREPARED

Read through the experiment before you start. Make sure you have all the ingredients and supplies.

. . . and so is the wizard's mother.

3. A CLEAN WIZARD IS A HAPPY WIZARD

Make sure your work area is clean before you start, and always clean up after yourself when you are finished. Wear an apron, wizard robe, or an old shirt so you don't get your clothes dirty, and keep a roll of paper towels or a damp dishcloth nearby so you can wipe up any spills.

4. A GOOD WIZARD WILL FOLLOW DIRECTIONS THE FIRST TIME . . . THEN EXPERIMENT *You don't want to turn yourself into a frog on the first try.*

In order to get the intended results of a potion you must be exact in your measurements and follow the directions carefully. After you have made the potion, we encourage you to try it again and experiment with it.

My uncle has his own rules for potion making.

1— The bigger the mess, the more success. Mom doesn't agree with this one.
2— All unmarked potion bottles contain extremely toxic poisons.
3— Always work with a friend. Then when something goes wrong, you have someone else to blame.

Contents

Chapter 1 — BASIC POTIONS FOR THE BEGINNING WIZARD

- **Love Potion** (Make two people fall in love) 2
- **Expanding Potion** (Keep mummies away) 6
- **Bogeyman Bomb Bag** (Scare away the bogeyman) 10
- **Monster Mushrooms** (Catch gnomes, sprites, or brownies) 14
- **Captured Dragon's Breath Potion** (Protect your home from dragons) 18
- **Enchanted Embers** (Protect yourself from any threat) 22
- **The Wishing Potion** (Grants a wish) 26

Chapter 2 — ADVANCED WIZARD POTIONS

- **Night Fairy Potion** (Attracts night fairies) 32
- **Zombie Fire** (Attracts zombies) 36
- **Goblin Goo** (Make troll snot) 40
- **Vamp Trap Potion** (Captures a vampire's powers) 44
- **Ghost Detecting Potion** (Tells you if a ghost is nearby) 48
- **Exploding Fog Fluid** (Sleeping potion for giants) 52
- **Bubbly Brew of Fire** (Breaks a witch's curse) 56

Chapter 3 — *CAUTION* HAIR-RAISING POTIONS FOR DARING WIZARDS

- **Essence of Death** (Removes and traps evil curses) 62
- **Monster Mists** (Overpower a werewolf) 66
- **Truth by Fire** (Indicates whether someone is lying or telling the truth) 70
- **Ultimate Power Potion** (Gives you power to control the elements: earth, wind, fire, and water) 74
- **Power Potion Paper** (Gives you any power you want) 78
- **Transfiguration Potion** (Turns people into animals) 82

Bonus Activity: Magical Stones 86
Potion Ingredients List 88

Chapter 1

BASIC POTIONS FOR THE BEGINNING WIZARD

Captured Dragon's Breath, pg 18

Love Potion

This potion has been around for centuries. It was invented by Ineta Hug, a castle maid who fell in love with Prince Will B. Luv. After Ineta gave the prince the potion, he fell in love with her and they lived happily ever after . . . until a witch turned him into a toad (see the transfiguration potion on page 82).

What You Will See

A liquid will burst with colors that magically move by themselves.

What It Is Used For

To make two people fall in love. It also makes a nice treat for a thirsty cat.

Ingredients

Spell Paper
1x2" piece of paper towel

Unicorn Milk
Milk 2% or Whole

Dragon Blood — 1 drop
Red food coloring

Toxic Fluid — 1 drop
Green food coloring

Supernatural Snot — 1 drop
Yellow food coloring

Troll Tears — 1 drop
Blue food coloring

Tubeworm Pus — 1 drop
Dawn liquid dish soap

Other Supplies

- Pen or marker
- Cotton swab
- Dinner plate
- Magic wand

Instructions

~ Time: 10 minutes ~

1. Prepare the **Spell Paper.** Use a pen or marker to draw a small heart in the middle of the paper and color it in. Be careful. A lopsided heart will result in a one-sided love for the couple. Now write the names of the two people whom you want to fall in love. Put one name on each side of the heart so that the heart is between them. Again, be careful. Make sure to spell the names correctly or something will go wrong with the potion. Now set the **Spell Paper** aside for later.

2. Pour some **Unicorn Milk** onto a dinner plate. Pour enough to completely cover the bottom of the plate to the depth of about ¼ inch.

3. Put in 1 drop of each of the following: **Dragon Blood, Toxic Fluid, Supernatural Snot,** and **Troll Tears.** Keep the drops close together, almost touching, in the center of the **Unicorn Milk.**

4. Put the first names of the two people whom you want to fall in love in the blank spaces below then repeat the words of the spell as you wave a wand over the potion:

> Boys have cooties, girls have fleas,
> Kissing is as gross as moldy cheese.
> But _____ and _____ their love shall be,
> So no more cooties, no more fleas.

5. Place 1 or 2 drops of **Tubeworm Pus** on the end of a cotton swab. Touch the tip of the cotton swab to the center of the **Unicorn Milk** between your drops of color. Do not stir. Just hold the cotton swab in the center of the plate for 10 seconds. You should see a burst of color.

6. Remove the cotton swab and watch the potion for about 30 seconds. If the colors continue to move on their own, then the potion has started to work.

7. Take the prepared **Spell Paper** and gently drop it onto the colored milk. Watch it carefully. If it floats for more than 5 seconds and absorbs some of the color, then it's likely that the couple will form a relationship. If one side of the **Spell Paper** sinks, then the person who's name is on that side is unwilling to let the love into their heart.

8. To seal the potion, use the cotton swab or your finger to move the **Spell Paper** around a little bit. Now pick up one corner of the paper and pull it around through the milk to swirl the colors even more. The prettier the pattern, the stronger the love will be.

=== More Fun ===

Cut a heart shape out from a piece of white copy paper or from a coffee filter then lightly touch the paper to the colored swirls in the potion. Do not push the paper down into the liquid. Just lightly touch it to the colors and then quickly remove it. Lay it somewhere to dry. Once it is dry, you can use a marker and write the names of the two people for whom the Love Potion was made for.

You really do have to be careful. I did this potion for my friend, Max, but I spelled a name wrong and he fell in love with the neighbor's dog!

Expanding Potion

This is an ancient potion. The directions for making it were first found in the tombs of Egyptian pharaohs. Until recently, the ingredients have been kept secret, but we have decided to reveal them here.

What You Will See

Pieces of a soap bar will expand and swell like a cloud of foam.

What It Is Used For

This potion is used to keep mummies away. It can also be used to double the contents of your soup pot when unexpected guests arrive.

Ingredients

	Pixie Soap
1 bar	Ivory soap bar

Who changed King Tut's diapers? His mummy! Ha,ha,ha!

Other Supplies

- Plate
- A microwave oven
- Knife

TOP SECRET INGREDIENTS LISTED BELOW (Shhhh . . . Don't tell anybody.)

Top Secret Ingredients

- 20 grams of Dead Man's Fat
- 15 drops of Troll Sweat (from the armpit)
- 2 Puffer Fish Eyes.

Please pretend that you did not read those ingredients. They must remain a secret because so many people have died while trying to make this potion. Why is it so dangerous? Trolls. DO NOT EVER try to collect sweat from a troll's armpit—they will kill you. We only listed the ingredients here because we have another, safer way for you to make it. Simply use Pixie Soap. Pixie Soap is already made with these 3 ingredients, and it can be found at any store that sells wizard supplies.

Adult Wizard Info

~ WARNING ~
This potion involves using a knife and a microwave oven.
It should only be done with the help of a grownup wizard.

Instructions

~ Time: 10 minutes ~

1. Get a bar of **Pixie Soap**. It must be fairly new. Old soap will not work.

2. Using the knife, carefully cut the bar into smaller pieces (about 1-inch squares).

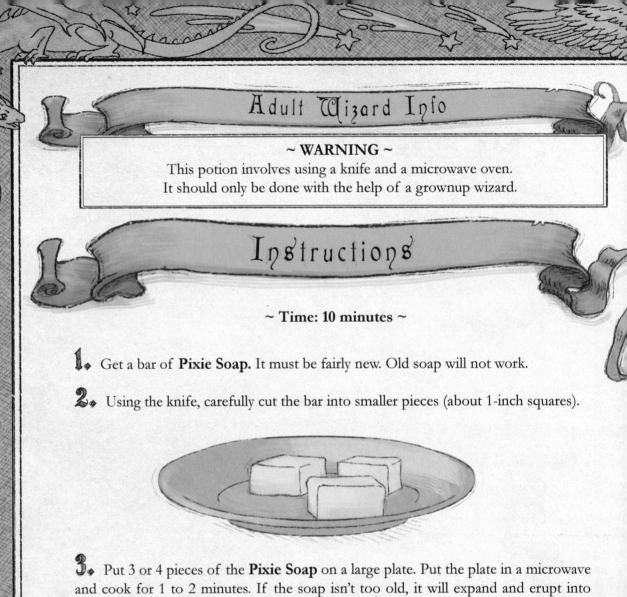

3. Put 3 or 4 pieces of the **Pixie Soap** on a large plate. Put the plate in a microwave and cook for 1 to 2 minutes. If the soap isn't too old, it will expand and erupt into puffy clouds. Watch it carefully, and when it looks like it has stopped growing, turn the microwave off.

4. Allow the soap to cool for a minute before touching. It will feel brittle and flaky.

5. Use as normal soap (it still has the same cleaning power as before), or see the following directions for how to keep a mummy away.

What do you call a mummy who eats cookies in bed?

A crummy mummy!

How to Keep a Mummy Away

Take your expanded Pixie Soap and scatter it around the area that you want to keep a mummy away from. We suggest that you scatter it all the way around the outside of your house. These small pieces of soap, are dangerous to mummies. If a mummy touches the soap he will vanish into thin air. All that will be left will be a pile of wrappings—which you can use as toilet paper.

My mom found out that I had 3 tuna fish sandwiches hidden under my bed. I didn't want to get in trouble, so I made this potion to keep my "mommy" away. It didn't work, but the soap did come in handy when she made me clean up that old, smelly tuna.

Bogeyman Bomb Bag

This potion was accidentally discovered by Anita Knapp. Anita never got any sleep. Her house was infested with Bogeymen, and they haunted her both night and day. Finally, Anita decided to move from her house. She packed a few potion ingredients to take with her, but she was too tired to do anything right. Standing in front of her potion cupboard, she dozed off and accidentally spilled Dragon Spit over her other ingredients, which caused a big explosion. The big bang scared away all of the Bogeymen, and Miss Anita Knapp was finally able to take a nap.

What You Will See
A small explosion in a bag.

What It Is Used For
This concoction will scare away the Bogeyman. It is also good for startling a sleeping cat.

Ingredients

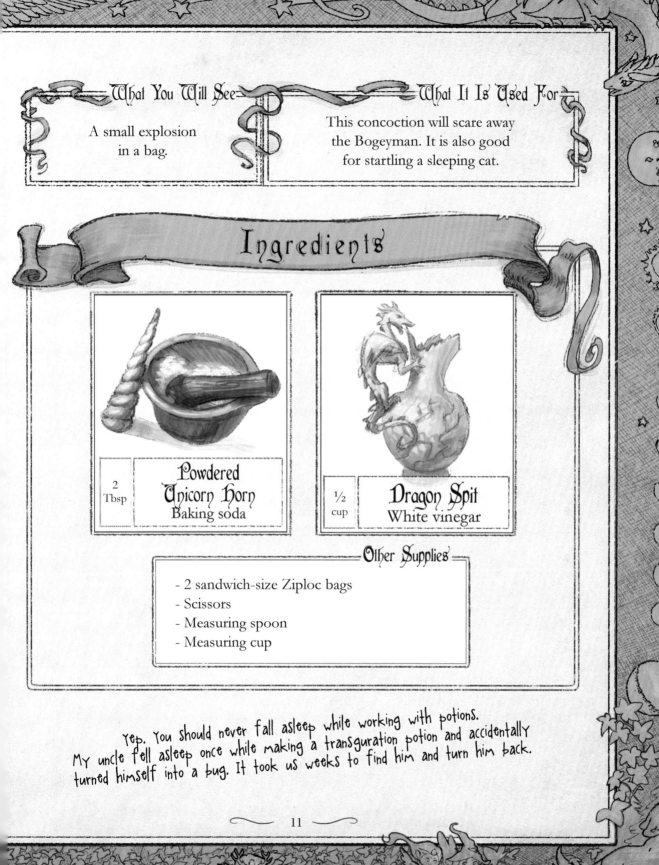

| 2 Tbsp | **Powdered Unicorn Horn** — Baking soda |
| ½ cup | **Dragon Spit** — White vinegar |

Other Supplies
- 2 sandwich-size Ziploc bags
- Scissors
- Measuring spoon
- Measuring cup

Yep. You should never fall asleep while working with potions. My uncle fell asleep once while making a transguration potion and accidentally turned himself into a bug. It took us weeks to find him and turn him back.

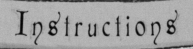

Instructions

~ Time: 10 minutes ~

1. Put 2 tablespoons of **Powdered Unicorn Horn** into a Ziploc sandwich-sized bag. Shake the bag so all of the powder falls into one corner.

2. Take another Ziploc bag and, using scissors, cut off the top 2 inches. This will remove the ziploc portion. Discard the top. In the newly, shortened bag, pour ½ cup **Dragon Spit.** You may need someone to hold the bag while you pour so it doesn't spill out.

3. Now gather the top of the bag together. Hold it tightly in one hand while you twist the bottom with the other hand. Twist it 3 or 4 times to make a tight pouch.

4. Set the pouch inside the other Ziploc bag in the opposite corner from the **Powdered Unicorn Horn.** The top of your pouch may untwist a little bit. This is all right as long as the liquid doesn't get out. Carefully seal the Ziploc bag shut. Double check it to make sure that it is indeed sealed.

5. Carefully take the bag outside to where you think Bogeymen might be coming into your home. You do not want the **Dragon Spit** to mix with the powder, so do not jostle the bag.

6. Once you have selected a spot, turn the bag upside down, letting the liquid run out of the pouch. Tip the bag so that the liquid runs into the powder. You may need to shake the bag once or twice then QUICKLY set the bag down and step back. The bag should expand like a balloon and explode open with a bang.

7. Say goodbye to the Bogeyman.

Who was the best dancer at the Halloween party?

The Bogeyman!

Why is the Bogeyman under your bed?

To tend to his bed bug farm.

Monster Mushrooms

Do you have a garden or flowerbed that is infested with pesky creatures such as brownies, sprites, and gnomes? This concoction will temporarily paralyze these annoying creatures so that you can capture them. If you don't want to catch these little creatures, you can always make this concoction and put it in your shoes to make your feet smell sweet.

What You Will See
Black mushroom-like objects will grow out of a powder.

What It Is Used For
To temporarily paralyze pesky creatures like brownies, sprites, and gnomes. It's also good for masking odors like stinky feet and wet dog.

Ingredients

2 tsp — Fairy Wing Powder
Powdered sugar

½ tsp — Powdered Unicorn Horn
Baking soda

Other Supplies

- Table salt or sand
- Lighter fluid
- Measuring spoons
- 2 bowls
- Long-handled lighter

What kind of bread do gnomes make sandwiches with? Shortbread of course! Ha,ha,ha! I told this to my friend, Max, and he laughed so hard, milk came out his nose!

Adult Wizard Info

~ WARNING ~
This potion involves using a lighter and should only
be done with the help of a grown-up wizard.

Instructions

~ Time: 15 minutes ~

1. In a small bowl put 2 teaspoons of **Fairy Wing Powder** and ½ teaspoon of **Powdered Unicorn Horn.** Mix well and set aside.

2. In the other bowl, create a dome of table salt or sand and press in the top to form a crater. It should resemble a small volcano.

3. Poor lighter fluid over the top of the salt (or sand) so that the entire crater is damp.

4. Place ½ of your powdered mixture in the crater.

5. Using a long-handled lighter, carefully light the powder on fire.

Black mushroom-like objects will begin to form and will continue to grow for several minutes until the fire goes out.

Catch pesky creatures by placing burnt remains of Monster Mushrooms around flowers or garden at night. The sweet marshmallow-like smell will attract them. Eating the burnt ashes will paralyze the small magical beings for several hours. By morning, there should be several creatures laying around and mumbling incoherently.

If I catch a gnome, then I'm going to put a petrifying spell on him and set him in my front yard. I hear gnomes make good lawn ornaments.

Captured Dragon's Breath Potion

Gone are the days when men would risk life and limb by going into a dragon's cave to capture a dragon's breath. With the development of this potion, it is easy to extract a dragon's breath from the creature's spit. Since dragon spit is available at any magic shop, you can make dragon's breath from the safety of your own home without ever facing a fiery beast.

What You Will See

A flame from a candle will be magically extinguished without blowing on it.

What It Is Used For

Putting out a candle using Captured Dragon's Breath will protect your home from dragons. It also comes in handy when an elderly person doesn't have the breath to blow out their own birthday candles.

This is Grandpa!

Ingredients

1 tsp — **Powdered Unicorn Horn** — Baking soda

2 ½ Tbsp — **Dragon Spit** — White vinegar

Other Supplies

- 1 extra-large cup or a pitcher (large enough to hold one quart)
- 1 or 2 small glasses
- Long-handled lighter
- Candles
- Glass jar with lid (optional for experiment)
- Measuring spoons

Adult Wizard Info

~ WARNING ~
This potion involves using a lighter and should only be done with the help of a grown-up wizard.

Instructions

~ Time: 10 minutes ~

1. With the help of an adult wizard, light a candle.

2. Add 1 teaspoon of **Powdered Unicorn Horn** to the extra-large cup or pitcher. The container must be large or else the breath will not be captured, but will flow out into the surrounding air and be lost.

3. Premeasure 2 ½ tablespoons of **Dragon Spit** into a small cup and then pour it all at once over the **Powdered Unicorn Horn**. This mixture will foam as the dragon's breath is released from the spit.

4. Dragon's breath is heavier than air, so it will stay in the container, however, you will not be able to see it. Put a hand over the container to prevent the dragon's breath from escaping. When the foam settles, pour the breath over the candle, but make sure you don't dip the container so far that the liquid comes out. The flame should appear to magically go out on its own.

Dragons are not very friendly and will usually stay away when they smell another dragon in the area. If there is a dragon living close to you, he will smell the dragon's breath you created and will stay away. If this doesn't work and a dragon starts attacking your house, you should run for your life.

Experiment

Dragon's breath cannot be seen, but it is still there. Try pouring the breath from your large container into a smaller, empty glass. Now pour the seemingly empty glass over the candle flame.

Try pouring the breath into an empty bottle and put the lid on it. The breath should keep so you can try to put a candle out with it later.

Try placing several small candles up a tiered platform (It should look like little stairs. You can stack books to make this). Make sure the candles line up with each other. Pour the dragon's breath over the top candle. The rest of the candles below should go out in turn as the dragon's breath invisibly flows down the steps.

I bet my uncle's breath smells worse than a dragon's.

Enchanted Embers

Enchanted Embers is a great spell which can be used to protect yourself from any evil creature. However, please be aware that this spell uses Curse Paper, which when lit on fire will blind a Cyclops. This was accidentally discovered when the famous Cyclops, Kent C. Strait, used this spell to protect himself from the sandman who was determined to sprinkle sand in his eye. Although the spell protected him from the sandman, Kent C. Strait went completely blind. Do not do the Enchanted Embers spell near any of your Cyclops friends.

What You Will See

A small object lit on fire will magically lift into the air and float up to the ceiling. The ash will then slowly float back down.

What It Is Used For

Protects a person from anything evil or threatening. It will also blind a Cyclops.

Ingredients

| 1 tea bag | **Curse Paper** Tea bag |

Other Supplies

- Scissors
- Pen
- Magic wand
- Long-handled lighter
- Non-flammable plate
- A cup with water

Adult Wizard Info

~ WARNING ~
This potion involves using a lighter and should only
be done with the help of a grown-up wizard.

Instructions

~ Time: 10 minutes ~

*The slightest breeze will affect this experiment, so it should be preformed indoors.
Make sure that you are well away from curtains or other flammable materials.*

1. Fill a cup halfway with water and set it aside. You will need this after you light the **Curse Paper** on fire.

2. Get the **Curse Paper**. **Curse Paper** is easy to find because it is commonly used to make tea bags. Any disposable tea bag should work (the kind with the string and the paper tag).

3. Using scissors, cut off the top of the tea bag to remove the staple and string.

4. Dump out the tea.

5. Unfold the tea bag to straighten it out. If the bag is not open on its ends, cut the end off so it can open like a tube.

6. Using a pen, write what you want to be protected from on the bag. This can be the name of a monster, your greatest fear, or a curse that you need protection from.

7. Use your fingers to open up bag to form a tube. You may need to crease the bag so it will stay open.

8. Stand the tube on its end on a flat, nonflammable plate (a dinner plate works perfectly) then wave a wand slowly over it 3 times as you repeat the words to the following spell:

By the light of this fire and the hair of my nose,
protect me from the top of my head, right down to my smelly toes.

9. Make sure that you are not directly under a ceiling light, near curtains, or other flammable objects. Use a long-handled lighter to ignite the top of the **Curse Paper** (tea bag).

10. As the fire burns down the **Curse Paper,** it will suddenly lift into the air and float toward the ceiling. Take the cup of water and hold it until the ash begins to float back down then catch the ash in the cup.

11. Swirl the ash around in the cup several times to complete the spell and then quickly pour it down the drain.

~ NOTE ~
Enchanted Embers will only protect you for 8 minutes.
After that you must repeat the spell or run for your life.

Make sure to catch the ash in the cup! Sam Sparmont (AKA Sammy-the-face-slammer) shot a blue curse at my face. I quickly did the Enchanted Embers spell, but I didn't catch the ash in the cup, and the curse hit me. My face was blue for a week. Everyone called me the walking blueberry!

Wishing Potion

This potion was invented by Ann Teak, the oldest woman living on the earth. At the time this book was published, she was celebrating her 342nd birthday. When Ann was younger, she thought she was the most beautiful woman alive. She never wanted to die because she didn't want to deprive the world of her beautiful face. So Ann invented this potion and wished that she would live forever. Her wish came true. She then made some alterations to the potion and made it public so that anyone could use it. You will note that the spell below refers to the person as an "ugly fellow." That is because Ann believed everyone was ugly when compared to her.

A Word of Warning:

It is not advised to wish to live forever. Although Ann Teak did get her wish, she continued to age and eventually her beauty gave way to the wrinkles of time. Ann Teak is truly antique and currently looks like a dried-up prune.

What You Will See

Little colored droplets streak down through a jar of liquid, forming amazing bursts of color.

What It Is Used For

This potion will grant a wish, or you can use it to celebrate the 4th of July if you can't afford fireworks.

Ingredients

2 Tbsp — Oil of the Salamander
Cooking oil

2 drops — Supernatural Snot
Yellow food coloring

2 drops — Toxic Fluid
Green food coloring

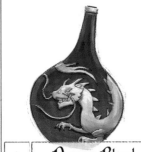

2 drops — Dragon Blood
Red food coloring

Water

Other Supplies

- 1 clear quart-sized jar
- 1 small container
- Fork
- Measuring spoon
- Magic wand

Instructions

~ Time: 10 minutes ~

1. Fill the clear quart-sized jar almost to the top with room-temperature **Water**.

2. In the small container, put 2 tablespoons of **Oil of Salamander**.

3. Add 2 drops of **Supernatural Snot,** 2 drops of **Toxic Fluid,** and 2 drops of **Dragon Blood** to the oil.

4. Use a fork to stir the oil, breaking up the drops of color. Do this by: quickly stirring 5 times to the right, then spinning yourself in a circle 3 times, then quickly stirring the oil 4 times to the left.

5. Make sure that most of the beads of color are floating in the oil and not stuck on the bottom of the container and then pour the oil mixture into the jar of water.

6. Wave the wand over the jar and repeat the following spell:

Streaks and drizzle of red, green, and yellow,
grant a wish to this poor, ugly fellow.

7. Now watch the liquid in the jar. As the drops of color begin to streak down through the water, make your wish.

~ RULES OF WISHING ~

1. You cannot wish for more wishes.

2. You cannot wish to bring someone back from the dead. (See how to attract zombies on page 36, or how to detect a ghost on page 48.)

3. You cannot wish for someone to fall in love with you. (See the Love Potion on page 2.)

I wished for Aunt Gertrude's food not to kill me. It must have worked. I had to eat her green-bean casserole last Sunday. It tasted like cat food, but I'm still alive.

Experiment

You cannot wish for more wishes, but you can try to trick the potion into giving you an extra wish. Follow the instructions on making the potion until you have made your wish and the colors have started to streak through the water. Now take a fork and put it down into the potion and begin to stir the water in a small circle. The colored streaks should appear to shrink backward and disappear. Remove the fork, repeat the words of the spell, and make a second wish. The colors should start streaking down from top of the water again and your second wish should be granted.

Wishes are unpredictable. Please note there are no guarantees or exchanges.

Chapter 2

ADVANCED WIZARD POTIONS

Night Fairy Potion, pg 32

Night Fairy Potion

This potion was developed by the famous elf princess, Constance Noring, who was fascinated with night fairies, but could never stay awake long enough to catch any of the little creatures. Although she set traps for the fairies, she couldn't help but fall asleep, and her loud snoring frightened away everything within earshot. So, Constance developed the Night Fairy Potion. This bubbly brew worked perfectly to attract night fairies, and Constance discovered that watching the bubbling liquid kept her awake long enough to actually catch the little creatures. After that there was no more snoring for Constance Noring.

What You Will See

A liquid that bubbles and fizzes and has colored blobs moving slowly through it.

What It Is Used For

This potion is used to attract night fairies. It also makes a good base for troll soup, or you can use it to polish furniture.

Ingredients

1–2 cups — Oil of the Salamander
Cooking oil

10 drops — Dragon Blood
Red food coloring

Fizzy Fizzing Fizzbee Tablets
Alka-Seltzer tablet

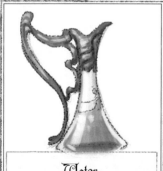

Water

Other Supplies

- 16-oz clear glass or plastic bottle (soda or water bottles will work)
- Funnel
- Measuring cups

Instructions

~ **Time: 10 minutes** ~

1. Fill the bottle ¼ of the way full with **Water.**

2. Use a funnel and slowly fill the rest of the bottle with **Oil of Salamander,** stopping about 1 inch from the top.

3. Wait a few minutes for the liquids to separate.

4. Add 10 drops of **Dragon's Blood.** (If you want another color you may use **Shifting Mists-blue,** or **Toxic Fluid-green.**)

5. Break a **Fizzy Fizzing Fizzbee Tablet** in half and drop 1 half into the bottle. It will sink to the bottom, and the bubbling will begin.

6. Continue adding **Fizzy Fizzing Fizzbee Tablets** as needed.

TO ATTRACT NIGHT FAIRIES, wait until after dark and then set the bottle on the head of a flashlight. The glow shining through the bubbling potion will attract the fairies. If this doesn't work, try clapping your hands together a few times. We don't know why, but fairies seem to like it when people clap.

I caught 2 night fairies!!!

Experiment

Try dropping in a Fizzy Fizzing Fizzbee Tablet
and then put the lid on.
Wait a minute and then remove the lid.
Try different oils such as mineral oil or baby oil.

~ NOTE ~

Night Fairy Potion can be stored for later use. It will last for several months as long as it is kept in a bottle and tightly capped. You may need to open it once in a while to release any pressure that has built up then just add Fizzing Fizzbee Tablets any time you want.

Zombie Fire

This potion was invented by Professor Willy Rott in 1879. Willy Rott was a well-known scientist, but after making this potion several times, he is now a well-known zombie. (He was not a very fast runner.)

What You Will See

A bright, mysterious green fire.

What It Is Used For

Green fire is used to attract zombies. It's also useful when having a weenie roast with trolls.

Who would do this?

Ingredients

1 tsp	**Scarab Beetle Powder** — Boric acid
2 Tbsp	**Zombie Acid** — Yellow bottle of Heet™

Other Supplies

- Spoon
- Glass jar or a disposable plastic cup
- Metal pan
  (brownie pan or cooking skillet works well)
- Long-handle lighter
- Heat-safe surface or hot pad

Adult Wizard Info

~ WARNING ~
This potion involves using a lighter. It will create a large flame
and should be done with the help of a grown-up wizard.

About this ingredient: BORIC ACID *is a relatively safe household chemical.
It is a white powder.*
Where to get it: *You can get boric acid in the pharmacy of most stores.
If they don't have it in stock, they can order it in for you. You can also buy Enoz Roach Away™, which is 99% boric acid and is sold with household insecticides.*

About this ingredient: HEET™ *GAS LINE ANTIFREEZE AND WATER REMOVER (the one in the yellow bottle) is mostly methanol or methyl alcohol.*
Where to get it: *Any supermarket, gas station, or auto supply store.
It is sold with automotive chemicals/additives.*

Instructions

~ Time: 10 minutes ~

Please note that this potion works best in the dark.

1. Select a location to make the potion. Outdoors may be best. There isn't a lot of smoke produced, but it may set off a smoke alarm if you're directly under it.

2. Pour 2 tablespoons of **Zombie Acid** into a glass jar or disposable plastic cup. Replace the cap on the original container of the **Zombie Acid** and place it a safe distance away from your work area.

3. Put 1 teaspoon of **Scarab Beetle Powder** into the liquid in the jar. Stir for 10 seconds. The powder may not dissolve all the way.

4. Set a metal pan on a heat-safe surface or hot pad, and put 1 teaspoon of the potion in it.

5. Turn out the lights (if you are inside) then stand back and use a long-handled lighter to ignite the liquid in the metal pan. The fire will burn a bright green.

6. The Zombie Fire will last for up to a minute before it goes out. After it has completely burned out, put another teaspoon of the potion on the tray and relight it. This can be repeated as many times as you like. Once you are done, allow the metal pan to cool for several minutes before touching it.

BE AWARE that constant relighting of this fire will attract zombies. If that is your plan, please be prepared to defend yourself.

Do this potion with a friend. Choose someone who runs slower than you do. To get away from zombies, all you have to do is to run faster than your friend.

— CAUTION —

SIDE EFFECTS

You may notice that you or those around you may start acting like zombies. This is a natural side effect from relighting a Zombie Fire several times. Don't worry; the effects are temporary and you will return to normal after a few minutes.

WARNING: If someone around you starts acting like a zombie, don't let them touch you, or the condition may pass to you as well.

We did this potion at school, and half my classmates turned into zombies, including Sara Sallis, the cutest girl in my class. To get away from her, I had to slam the door into her.

She still looked smashing. Get it? Smashing!

Goblin Goo

Many people still prefer troll snot over Goblin Goo; however it is no longer legal to collect snot from trolls. Therefore Goblin Goo should be used instead. Please note that anyone found picking a troll's nose or collecting troll snot will face a board of inquiries lead by the head troll, Mr. Iva P. Brain.

What You Will See

Green slime.

What It Is Used For

A common substitution for troll snot. Goblin Goo is also used in ghost traps and is a favorite toy for young goblins.

Ingredients

1 tsp	**Octopus Powder** — Borax
1/2 cup	**Pond Slime** — Clear Elmer's glue
6 drops	**Toxic Fluid** — Green food coloring
1 1/2 cup	**Water**

Other Supplies

- 2 containers
- Stirring spoon
- Measuring spoon
- Measuring cup

Instructions

~ Time: 20 minutes ~

1. In one container, mix 1 teaspoon of **Octopus Powder** in 1 cup of **Water** and stir until the powder is dissolved.

2. In the other container, mix ½ cup **Pond Slime** with ½ cup water. Add 6 drops of **Toxic Fluid** and stir until smooth.

3. Now to combine the two solutions. Pour the **Pond Slime** solution into the other solution. The **Pond Slime** will begin to congeal into Goo immediately.

4. Mix the solution up as much as you can, then take the Goo out of the liquid and finish mixing it by hand.

5. Now play with it as if it is troll snot or put it in any standard ghost trap. To store your Goblin Goo, keep it in a plastic bag or sealed in a small container.

> Do not make Goblin Goo and take it to school. If you do, don't put it on your friend's lunch tray when his back is turned. It's not a good idea. Not that I know. I'm just sayin.

=== More Fun ===

Try adding glow-in-the-dark paint to your Goblin Goo.

Vamp Trap Potion

Do you suspect a vampire is living in your neighborhood? Tired of wearing smelly cloves of garlic around your neck? The Vamp Trap Potion is the perfect solution. This potion will render any vampire powerless so you can remain bite free. Easy to use, this potion will affect any vampire within a 5-mile radius by making them sparkle for a few moments as it draws their powers into the mixture. Add a little Dragon Spit to trap the powers in the potion forever. There is no better way to befuddle a bloodsucker.

What do you call a dog owned by a vampire?

A blood hound.

What must you do to learn more about a vampire?

Join their fang club.

What did my friend say when I told him my vampire jokes?

They really suck.

What You Will See
This bright-yellow liquid will suddenly turn red as it dances and shimmers across a plate, moving as if it were alive.

What It Is Used For
This potion will make every evil vampire sparkle. It then turns them harmless by absorbing their powers.

Ingredients

1 tsp **Sneezewort Powder**
Turmeric powder

¼ cup **Death Tincture**
70% or higher Rubbing alcohol

1 ½ tsp **Powdered Unicorn Horn**
Baking soda

2 ½ Tbsp **Dragon Spit**
White vinegar

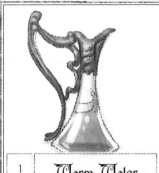

1 cup **Warm Water**

Other Supplies
- Paper towels
- Tray or cookie sheet
- 3 paper cups or disposable plastic cups
- Rubber band (optional)
- Scissors
- 1 sturdy paper plate
- Measuring spoons
- Measuring cups

How do you join a vampire's fan-g club?
Send your name, address, and blood type.

Adult Wizard Info

~ WARNING ~
Sneezewort Powder will stain. It is advised to wear gloves, unless you don't mind yellow fingers.

Instructions

~ Time: 30 minutes ~

1. Prepare your work surface. Lay several paper towels on a tray or cookie sheet. This will catch any spills. **Sneezewort Powder** will stain most surfaces, so it is important to do all of your work on the covered tray.

2. In a disposable cup, mix 1 teaspoon **Sneezewort Powder** in ¼ cup **Death Tincture.** Stir thoroughly.

3. Now to strain the solution. Put a paper towel over another cup, pushing it down inside to create a pocket. You may need to secure the paper towel with a rubber band. Slowly pour your solution into the paper towel to strain it.

4. Using scissors, cut 2 or 3 strips from another paper towel. Make the strips about 1 inch wide. Dip them partway into the strained solution, then lay them aside. These will be used later for experiments.

5. In another cup, mix 1½ teaspoons of **Powdered Unicorn Horn** with 1 cup of **Warm Water** and stir thoroughly.

6. Select a sturdy paper plate with a lip around the edge. A white one is ideal so you can see the color change in the solution. Pour about ¼ cup of the **Unicorn Horn** solution in it. The amount doesn't need to be precise as long as you cover the bottom of the plate and still have room so more liquid can be added.

7. Pour the **Sneezewort** solution into the liquid on the plate. DO THIS SLOWLY—A FEW DROPS AT A TIME. Wait several seconds between each pour, taking time to observe the reaction in the liquid. It should turn a shimmering red color and dance and across the plate. This is a sign that it is drawing the vampire's power into it.

This is fun! Somewhere there is a vampire sparkling right now!

8. As you continue to add the **Sneezewort** solution, the liquid in the saucer will gradually get darker showing that it is drawing more vampire power into it. The liquid may act strange, moving and shimmering as if it is alive.

9. When you have poured all of the **Sneezewort** solution into it, premeasure 2 ½ tablespoons of **Dragon Spit** into a cup then pour it all at once into the solution on the paper plate. It should fizz a little, which indicates that the power is trapped. Now you may discard it by carefully pouring it down the drain.

=== Experiment ===

You may use the yellow-dyed strips of paper towel to see if any vampires are in your neighborhood.

Place a drop of your Unicorn Horn solution onto a yellow part of a paper towel strip. If it turns a dark red, then there is a vampire within 5 miles of you. Now place a drop of **Dragon Spit** onto the red spot to see what it does. You can repeat the process as many times as you want to.

Ghost Detecting Potion

This potion was accidentally discovered by Ted N. Buried, who was a great scientist while alive, but an even greater one after he died. This particular potion was discovered when the ghost of Ted N. Buried was trying to come up with a new toilet cleaner but noticed that the liquid changed colors every time he got close to it. The potion is now commonly used to detect when a ghost is near.

What You Will See

A light-colored liquid will magically go dark for a few seconds and then go back to its original color.

What It Is Used For

Tells you when a ghost is nearby. It also makes a good toilet bowl cleaner.

Ingredients

1 **Flickerboom Tablets**
Glucose tablets or 1 ½ tsp of dextrose powder

WEAR GLOVES

½ Tbsp **Ground Snake Fang**
Sodium hydroxide or lye

5 drops **Shifting Mists**
Methylene blue

1 1/3 cup **Water**

Other Supplies

- 16 ounce CLEAR bottle with a tight fitting lid
- Plastic bag and rolling pin or mortar and pestle
- Eyedropper
- Measuring cups
- Measuring spoons

Adult Wizard Info

~ WARNING ~
Ground Snake Fang can irritate the skin. Please wear gloves.

About this ingredient: *SODIUM HYDROXIDE is commonly known as "lye," which is the active ingredient in lye soap. It is also known as "Caustic Soda."*

Where to get it: *On the internet (Amazon, eBay, and other sites). Chemical suppliers. Some hardware stores as a drain opener. (Lowe's usually carries "Crystal Drain Opener" made by Roebic Laboatories, which is 100% Sodium Hydroxide.)* SODIUM HYDROXIDE will irritate your skin, so wear gloves. If it touches you, wash thoroughly with soap and water.*

About this ingredient: *METHYLENE BLUE is commonly used as a medication for fish aquariums.*

Where to get it: *On the internet (Amazon, eBay, and other sites). Some stores that sell fish and aquarium supplies.*

Instructions

~ Time: 15 minutes ~

1. Pour 1 1/3 cups of **Water** into the bottle.

2. Using the plastic bag and rolling pin, or the mortar and pestle, pound 1 **Flickerbloom Tablet** into a fine powder and add it to the water in the bottle.

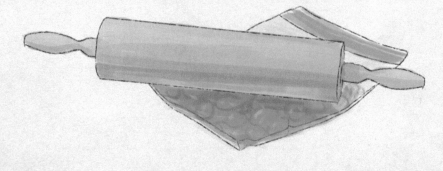

3. Add ½ tablespoon **Ground Snake Fang.** Swirl or shake the liquid to mix it.

4. Use an eyedropper to add 5 drops of **Shifting Mists** then let it settle. The liquid should be a light gray color with a thin layer of blue on the surface.

5. Cap the bottle tightly. To know if there is a ghost nearby, shake the bottle vigorously and observe the color change. It will only change colors for a few minutes, and then it will go back to its original color.

~ INDICATOR ~

NO CHANGE IN COLOR = No ghosts are present.

LIGHT GRAY = A ghost is in the room, but is not very close to you.

DARK BLUE OR MIDNIGHT BLUE = A ghost is standing right next to you.

WHAT TO DO IF A GHOST IS IN THE ROOM

1. Don't panic. Ghosts are usually harmless.

2. If the ghost is visible, calmly observe them to see what they do.

3. Try to start up a friendly conversation with the ghost.

Ignore all this and run like crazy!

Exploding Fog Fluid

Yeah, I'd draw mustaches or sleeping giants too! Maybe this guy is related to me!

This potion was made famous by Mr. Darren Deeds, who made a sport of puttin[g] Exploding Fog Fluid into the drinks of giants and then drawing mustaches on their faces onc[e] they were asleep.

Although this can be a fun pastime, please note that it's dangerous. As Mr. Darren Deeds ca[n] attest to, it's not very fun when a giant refuses to drink the potion and decides to sit on you instea[d].

What You Will See

A liquid that erupts into expanding foam.

What It Is Used For

Exploding Fog Fluid makes a great sleeping potion for giants. It is also effective in unclogging ogre hair from drains.

Ingredients

½ cup — **Dragon Acid Extra Strength**
20-volume hydrogen peroxide liquid

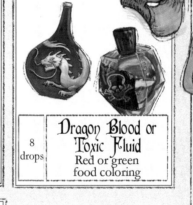

1 Tbsp — **Tubeworm Pus**
Dawn liquid dish soap

8 drops — **Dragon Blood or Toxic Fluid**
Red or green food coloring

1 Tbsp — **Crumb of Alligator Skin**
Packet of yeast

3 Tbsp — **Water**

Other Supplies

- 16-ounce plastic soda bottle
- Funnel
- Small cup
- Stirring spoon
- Measuring spoons
- Measuring cup
- Tray

Adult Wizard Info

~ WARNING ~
Dragon Acid – Extra Strength can cause eye and skin irritation. Please wear safety goggles and gloves.

About this ingredient: *20 VOLUME HYDROGEN PEROXIDE LIQUID is a 6% solution of hydrogen peroxide.*

Where to get it: *Beauty supply stores or hair salons. (Make sure to get the liquid and not the gel.)*

Instructions

~ Time: 10 minutes ~

1. Wearing gloves and goggles, use a funnel and carefully pour the **Dragon Acid** – Extra Strength into the plastic soda bottle.

2. Add 8 drops of **Dragon Blood** or **Toxic Fluid**.

3. Add 1 tablespoon of **Tubeworm Pus** into the bottle and swirl the bottle around to mix it.

4. In a separate small cup, combine 3 tablespoons of **Warm Water** with 1 tablespoon of **Crumbs of Alligator Skin.** Stir this mixture for 30 seconds.

5. Set the plastic soda bottle on a tray then use a funnel to pour the **Alligator skin** mixture into the bottle. It should begin to foam.

6. To put a giant to sleep, take the remains of the potion outside and set them close to where you suspect a giant lives. Wave your wand over it in a swooping motion while saying the following spell:

> *Potion of power*
> *potion of might*
> *make this giant*
> *sleep tonight*

If the potion is gone the next day, you know that the giant drank it and had a good night sleep.

Bubbly Brew of Fire

Is your neighbor a menacing witch who likes to curse you? Do her pesky crows hang out in your trees and leave bird droppings in in your yard? Now you can create some magic of your own and bewitch that troublesome witch or any other annoying neighbor or careless crow. The Bubbly Brew of Fire Potion is packed with the perfect power to break any witch's curse and repel her pesky crows.

What You Will See

A ball of fire will erupt from a bowl of sudsy bubbles.

What It Is Used For

This potion is used to break a witch's curse. It is also good for scaring crows out of trees.

Ingredients

Spell Paper	Tubeworm Pus	Water
1x2" piece of paper towel	25 drops — Dawn liquid dish soap	

Other Supplies

- Large glass bowl
- Air from an empty aerosol spray can
- Pen
- Long-handled lighter
- A paper with the words of the curse that you are trying to break.

Adult Wizard Info

~ WARNING ~
This potion involves using a lighter and will create a large flame, and should only be done with the help of a grown-up wizard.

About this ingredient: *The EMPTY AEROSOL CAN is needed for the remaining flammable propellant (air) which is left in the spray can.*

Where to get it: *Examples of empty spray cans to use: spray paint, hairspray, canned air, antiperspirant, and bug spray. Anything that contains a flammable propellant. (In this experiment, we will keep the can away from the flame for safety reasons.)*

Instructions

~ Time: 10 minutes ~

1. Tear off a small piece of **Spell Paper** and write the words "FLACTUS FLAMMAS" on it. This is a simple spell to create fire.

2. Fill a large glass bowl with **Warm Water.**

3. Gently set the prepared **Spell Paper** on top of the water, then stir 3 times and push the paper down until it sinks.

4. Add 25 drops of **Tubeworm Pus** and gently mix it with your hand. (Rinse your hand off afterwards.)

5. Put air into the potion. Do this by tipping an empty aerosol can upside down. Put the nozzle down in the water, and spray it. Even though it is empty, there should still be some air left in the can and it will spray into the water. Bubbles should begin to form on top of your potion.

6. Remove the can and set it a safe distance away from the potion. If you are trying to break a witch's curse, then say the words of the curse backwards. For example if someone used this curse on you:

I curse you now to fulfill my wishes
You shall smell like stinky fishes.

Then you would repeat the words in this order:

Fishes stinky like smell shall you
Wishes my fulfill to now you curse I.

7. Now to light the potion on fire. Stand back (especially if you have bushy eyebrows that you don't want singed). Use a long-handled lighter and ignite the bubbles.

This potion will not work on this curse: "Says master what do you do? You do what master says." Because this curse says the same thing forward and backward!

Ultimate Power Potion, pg 74
Transfiguration Potion, pg 82

Essence of Death

In 1904, an unknown sorcerer cast a terrible curse on the famous Chinese wizard, Hu Flung Pu. The curse would have killed Mr. Pu if he had not used this potion to save himself. When Hu Flung Pu was done with the potion, he threw it outside on the ground and discovered that by doing so, it summoned the Grim Reaper of Rodents. It would have been a pleasant discovery except the Grim Reaper took Mr. Pu's pet hamster as payment for their meeting.

What You Will See
A light-colored liquid will magically turn black all by itself.

What It Is Used For
Essence of Death is used to remove dark and evil curses. It is also used to summon the Grim Reaper of Rodents, or it can be used to clean your cauldron.

Ingredients

 **2 Orange Thistle Tablets** Vitamin C tablets	 **1 tsp Vampire Venom** Liquid iodine solution	 **1 Tbsp Dragon Stomach Acid** 3 % Hydrogen peroxide
 **½ tsp Spirit Dust** Cornstarch	**¾ cup Water**	**Other Supplies** - 3 CLEAR glass containers (half-pint jars or juice glasses will work) - Plastic bag and rolling pin or mortar and pestle - Measuring spoons - Measuring cups

Adult Wizard Info

~ WARNING ~

Vampire Venom will stain just about anything it touches and it can be hazardous. Dragon Stomach Acid can cause eye and skin irritation—safety goggles and gloves are advised.

Instructions

~ Time: 20 minutes ~

1. Set out the 3 CLEAR containers and label them #1, #2, and #3. You will be making 3 solutions.

2. Grind 2 **Orange-Thistle Tablets** into a fine powder. You can do this by placing the tablets in a plastic bag and crushing them with a rolling pin or using a mortar and pestle.

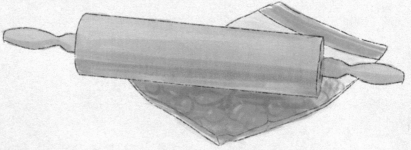

3. Put the powder into your into glass #1 and add ¼ cup of **Warm Water.** Stir for at least 30 seconds.

4. In glass #2 put ¼ cup of **Warm Water** and add 1 teaspoon of **Vampire Venom**.

5. In glass #3 mix ¼ cup of **Warm Water**, 1 tablespoon of 3% **Dragon Stomach Acid**, and ½ teaspoon of **Spirit Dust**.

6. Now to combine the solutions: Take 1 teaspoon of the solution from glass #1 and add it into glass #2. Notice how the dark liquid dissipates into a lighter color.

7. Now poor all of glass #2 into glass #3. Stir for 30 seconds, then set a timer for 3 minutes.

8. Now stare at the potion. Think about the curse that you're trying to get rid of. Don't take your eyes off of the liquid in the glass or you will miss the magic! Before the timer rings at the end of the 3 minutes, the potion should absorb the curse and turn black.

Try saying this 10 times really fast as you wait for the potion to turn black:

A cackling charmer
cast and conjured
in his cauldron
to catch a curse.

= Experiment =

Take 1–2 teaspoons of the leftover liquid from the fist solution and add it to the black liquid to see if it will turn lighter in color and then black again.

Try replacing the **Spirit Dust** (cornstarch) with ½ teaspoon of **Spirit Fluid** (liquid starch) and see if there is any difference in the reaction.

**AFTER THE POTION TURNS BLACK,
USE IT TO SUMMON THE GRIM REAPER OF RODENTS.**

Take the black potion outside and pour it onto the ground in a circle. The Grim Reaper will appear in the center (if he isn't busy collecting the souls of departed rodents). He is only 6 inches tall, so be careful not to step on him. It is also polite to give him a piece of cheese as a way of thanking him for coming.

Monster Mist

In the early 1900s there was a rash of unsolved crimes in a small town in North America. These were serious crimes that included robberies, break-ins, and even the theft of a picnic basket, which was taken in broad daylight in a crowded park. Mothers even reported ice cream being stolen right out of their children's hands. The crimes continued for months because no one was able to identify the thief as he always disappeared in a cloud of smoke. It was later discovered that the cloud of smoke was in fact, Monster Mists, and the thief was identified as the young wizard Ken U. Seemee. It is now illegal to use Monster Mists as a way of escaping crime scenes, although many young wizards still use it to sneak out of a classroom when the teacher announces a pop quiz.

What You Will See
A cake-like lump that makes A LOT of smoke when lit on fire.

What It Is Used For
Overpowers werewolves, and turns them human. Or you can use it as a smokescreen when you need to slip away unnoticed.

Ingredients

| 5 Tbsp | **Corpse Dust** — Stump remover or Potassium nitrate |

| 3 Tbsp | **Ground Dragon Egg** — Sugar |

| 1 Tbsp | **Powdered Unicorn Horn** — Baking soda |

Other Supplies
- Tinfoil
- 2 spoons
- Measuring spoons
- Pan
- Long-handled lighter

Adult Wizard Info

~ WARNING ~
This potion involves cooking on the stove and using a match or a lighter, and should only be done with the help of a grownup wizard.

About this ingredient: POTASSIUM NITRATE *is the active ingredient in stump remover. It is also known as "saltpeter".*

Where to get it: *Stump remover can be found in various stores. (Home Depot and Lowe's both carry "Spectracide Stump Remover.") If choosing a different brand, make sure that it says "contains potassium nitrate" somewhere on the label, which may be listed under "warnings." Please note that smaller granules work a little better than the larger bead form. You can also buy it online. (Amazon, eBay, and other sites). Search "Potassium Nitrate."*

Instructions

~ Time: 20 minutes ~

1. Prepare a place to put your hot mixture by laying out a piece of tin foil. The mixture will harden quickly, so you will want this ready ahead of time.

2. In a pan over medium heat, mix 5 tablespoons of **Corpse Dust** and 3 tablespoons of **Ground Dragon Egg**.

3. Stir until it melts. This will take several minutes and it will look something like creamy peanut butter. (If your **Corpse Dust** was in bead form, then it will look lumpy because the beads do not melt.)

4. Quickly remove it from the heat and stir in 1 tablespoon of **Powdered Unicorn Horn.** The concoction will turn a brown color and expand almost like foam.

5. Scoop large spoonfuls of the mixture onto the tinfoil leaving space between each lump. It is very sticky so you will need two spoons—one to scoop the mixture and the other to scrape it off and onto the tinfoil. ***Do not touch the mixture—it will burn you***

6. Let the lumps cool completely, then tear the tinfoil to separate them.

~ HOW TO USE MONSTER MIST ~
The mixture is ready to be used after it has cooled and hardened.

~ CAUTION ~
DO NOT USE INDOORS!

Take the cooled lumps of Monster Mist outside to a location known to have werewolves. Set the lumps on the ground where the werewolves are expected to pass. You may have to lure the creatures to you by baiting them. Suitable bait would be: rotten eggs, raw meat, or young children (a cousin or an annoying brother or sister will work). When you think that the werewolves are nearby, light the lumps of Monster Mist on fire with a long-handled lighter. When the smoke touches a werewolf, they will instantly be overpowered and will turn into their human form.

Please Note

Monster Mist does not work on Ghouls, Ghosts, Vampires, Zombies, Leprechauns, or annoying relatives.

I tried this on my Aunt. She has hair growing out of her chin, and she looks like a werewolf.

Truth by Fire Potion

You will need a small article of clothing from the person you are testing. A sock is ideal. Select one that doesn't smell too stinky—for your own sake.

Why the sock? Not long ago it was common to use all kinds of clothing for this potion—shirt, jacket, pants, etc. However, the Wizarding Board of Potion Management made a rule that only socks can be used from now on. Apparently a young man, Josh K. Ding, who worked in the Potions Management Department, loved to tease and joke with everyone. No one ever knew if he was telling the truth or lying. Josh had this potion used on him so many times that all of his clothes were set on fire. When Josh showed up to work in his underwear because he had no clothes, the Wizarding Board quickly made the rule that only socks can be used.

I use this potion every time my mom makes me fold laundry.
It's a great way to get rid of the mismatched socks!

What You Will See

Light someone's sock on fire. Despite the flames, it will not burn.

What It Is Used For

This potion is used to tell whether or not a person is lying. It's also just plain fun to light your friend's sock on fire.

Ingredients

| 1/4 cup | Death Tincture
70% or higher Rubbing alcohol | a pinch | Scarab Beetle Powder
Boric acid | 1/4 cup | Water |

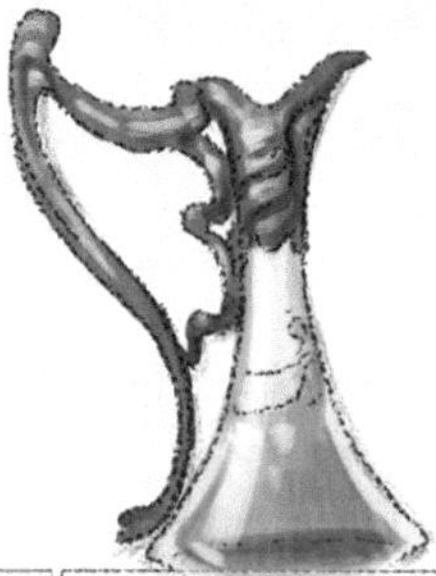

Table salt can be used if you don't have boric acid.

Other Supplies

- Sock
- Tongs
- Measuring cup
- Bowl
- Long-handled lighter
- Stirring spoon

Adult Wizard Info

~ WARNING ~
This potion involves using a lighter and will create a large flame. Use tongs to hold the object you are lighting on fire, and do this over a sink. This potion should be done with the help of a grown-up wizard.

About this ingredient: *BORIC ACID is a relatively safe household chemical. It is a white powder.*

Where to get it: *You can get boric acid in the pharmacy of most stores. If they don't have it in stock, they can order it in for you. You can also buy Enoz Roach Away™, which is 99% boric acid, and is sold with household insecticides.*

Instructions

~ Time: 15 minutes ~

1. Select a person you want to test for truthfulness and get one of their socks.

2. Make the potion by mixing ¼ cup **Death Tincture** with ¼ cup **Water,** and add a pinch of **Scarab Beetle Powder.** Mix it with a spoon.

3. Soak the sock in the potion until it is wet all the way through.

4. Using tongs, have the owner pick up his or her sock, and let any excess liquid drain.

5. Have the person move the wet sock away from the potion and hold the sock over a sink.

6. Ask the person a question and use a long-handled lighter to light the sock on fire. As the person answers the question, observe the flame to see if the person is telling the truth or lying. Let the fire go out on its own.

Truth or Lie

IF THE PERSON IS TELLING THE TRUTH: The fire will pulsate like a beating heart and will not consume or burn the object.

IF THE PERSON IS TELLING A LIE: The fire will go out quickly, or it will burn the object.

More Fun

Test yourself for truthfulness. Write the name of a person you love on a piece of paper, or write down your greatest desire. Soak the paper in the potion solution. Follow the safety procedures above. Hold the paper with tongs, then light it on fire. If the paper is not consumed and burned by the flames, then what you wrote is true.

Ultimate Power Potion

Please use extreme caution when using this potion. History is dotted with stories of tragedies and disasters that happened because of the misuse of this potion. For example: Remember Mrs. Scowers who almost drowned while cleaning her shower when she attempted to control the water. Another example is Doug D. Grave who was nearly buried alive when he tried to control the earth in order to save time planting his garden. It is even rumored that Dorthy was trying to control the wind when she was taken to the Land of Oz.

Special note: *Please refrain from creating tornadoes in the mountains or snow blizzards in the desert.*

I would also advise not to create a rainstorm during your cousin's outdoor wedding. At least don't get caugh

What You Will See
A blue liquid that can turn pink, purple or green.

What It Is Used For
This potion will allow you to control the elements: Water, Earth, Fire, and Wind. It can also be used to remove the smell in a gym locker.

Ingredients

Amount	Ingredient	Real Ingredient
2 Tbsp	Fire Fluid	Lemon juice
1	Red Cabbage	
1 tsp	Graveyard Dust	Cream of tartar
2 tsp	Powdered Unicorn Horn	Baking soda
1 Tbsp	Dragon Spit	White vinegar
1 Tbsp	Ogre Sweat (WEAR GLOVES)	Household ammonia
	Water	

Other Supplies
- Blender
- Strainer
- Pitcher or large container
- Tray (to catch spills)
- 4 small, CLEAR glass containers (half-pint jars or juice glasses will work)
- 1 small container
- Measuring spoons
- Stirring spoon
- Measuring cup
- Paper and pen to make labels

Adult Wizard Info

~ WARNING ~
This potion contains Ogre Sweat, which is toxic and has a horrible smell. Try not to breath it in. It can also cause eye and skin irritation—safety goggles are gloves are advised.

Instructions

~ Time: 20 minutes ~

1. Pull several large leaves off of the **Red Cabbage** and tear them into small pieces until you have about 2 cups. Place the cabbage in a blender, cover it with hot **Water,** and blend it.

2. Poor the liquid through a strainer into a pitcher or another large container. The liquid should be a dark purple-bluish color.

3. Set 4 clear glass containers on a tray. Pour ¼ cup of the dark cabbage liquid into each container then add ¾ cup **Water** to each of them. Give each container a label: Water, Earth, Fire, and Wind. We will be making 4 solutions, and each will need to be a different color.

WATER

In a small container, mix 1 teaspoon of **Powdered Unicorn Horn** with 2 tablespoons of **Water.** Pour the mixture into the glass labeled "Water." The solution should turn blue.

EARTH

Add 1 tablespoon of **Ogre Sweat** into the glass labeled Earth. The solution should turn green.

FIRE

Add 1 tablespoon of **Fire Fluid** into the glass labeled Fire. The solution should turn bright pink.

WIND

(Make sure you mix this solution on a tray.)

Add 1 tablespoon of **Fire Fluid** into the glass labeled wind. The solution should turn pink. Now add the following: 1 teaspoon **Powdered Unicorn Horn**, 1 teaspoon **Graveyard Dust**, and 1 tablespoon **Dragon Spit**. The solution will foam and should turn purple.

To control one of elements, select one of the potion solutions (Earth, Wind, Fire, or Water). Hold it up in the air as you repeat the words to the matching spell.

Power of the elements I command
*Essence of **Water** I hold in my hand*
Rain and sea and even ocean
I now control through this potion

Power of the elements I command
*Essence of **Earth** I hold in my hand*
Darkest dirt and lightest sand
You will do as I command

Power of the elements I command
*Essence of **Fire** I hold in my hand*
Whatever can flicker into flame
Do my will as I claim

Power of the elements I command
*Essence of **Wind** I hold in my hand*
From large tornado to little breeze
Do my will, I ask thee please

Power Potion Paper

The potion used to make Power Paper has a dark history—really dark. It started in the dark ages when a little boy was born at night, just as the last log on the fire burned out. It was dark. This little boy was named Benton D. Struction. As he grew up, he set a goal to become the darkest, most evil wizard known in history. No one could fear a wizard named Benton, so he changed his name several times. At one point he was known as Foster

Dennis Peeding Bullet. After so many names, no one knew what to call him so we will refer to him as You Know Who. You Know Who nearly achieved his goal of becoming the most evil wizard alive when he figured out how to trap superpowers inside a paper and then later absorb that power into his body. He nearly took over the world, but in the end was defeated by a paper cut.

What You Will See

You will make special paper and write on it with magical ink. A dark red mark on this paper will mysteriously disappear when you press your thumb to it.

What It Is Used For

This paper will allow you to absorb any super power you desire. It can also be used to make greeting cards for vampires.

Ingredients

1 tsp **Sneezewort Powder**
Turmeric powder

¼ cup **Death Tincture**
70% or higher Rubbing alcohol

1 tsp **Powdered Unicorn Horn**
Baking soda

½ cup **Warm Water**

Other Supplies

- Old newspaper
- Tray or cookie sheet
- 2 paper cups or disposable plastic cups
- Paper towel
- Scissors
- Foam brush
- Hair dryer (optional)
- White cardstock
- Stirring spoons
- Measuring spoons
- Measuring cups

Adult Wizard Info

~ WARNING ~
Sneezewort Powder will stain. It is advised to wear gloves, unless you don't mind yellow fingers.

Instructions

~ Time: 30 minutes ~

1. Prepare your work surface. Lay an old newspaper on a tray or cookie sheet. This will catch any spills. **Sneezewort Powder** will stain most surfaces so it is important to do all of your work on the tray.

2. In a disposable cup, mix 1 teaspoon **Sneezewort Powder** in ¼ cup **Death Tincture.** Stir thoroughly then let it settle while you do the next step.

3. Make the Power Paper. Cut the white cardstock in half to make 2 smaller cards. Lay these cards on the newspaper-covered tray and use a foam brush to paint the paper with the yellow Sneezwort solution. You will have enough solution to cover several sheets of paper so make as much Power Paper as you want. You can save the extra paper to use later. Let the paper dry for about 10 minutes, or use a hair dryer on a low setting to help it dry faster.

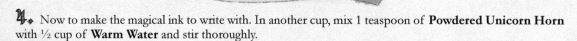

4. Now to make the magical ink to write with. In another cup, mix 1 teaspoon of **Powdered Unicorn Horn** with ½ cup of **Warm Water** and stir thoroughly.

5. If your Power Paper is dry use a paper towel and brush off any grains of **Sneezewart Powder** that might have stuck to the paper. Throw away the paper towel and the old newspaper, but keep the tray to work in.

6. Select a piece of dry Power Paper and place it on the tray. You will be using your finger to write with so do not wear gloves for this part. Say the following words:

With my finger, I now write
Grant me power on this night.

7. Now dip your finger into the magic ink and write the name of the superpower on the Power Paper. You may have to redip your finger in the ink solution several times.

8. Sign the Power Paper by putting a thumbprint on it. Dip your thumb into the magic ink and then press it to the Power Paper.

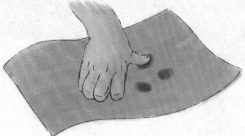

9. Put the Power Paper somewhere where no one can see it. This could be in a closet or under your bed. Let it set for about 20 minutes. During that time the Power Paper will draw the superpower into it and trap it there. Look at the paper after 20 minutes. If the writing and your thumbprint have turned a dark red, then the potion has worked and the superpower is trapped inside the paper.

Continue to part 2 to learn how to get the superpower from the Power Paper.

Part 2

ABSORBING POWERS FROM THE POWER PAPER

~ INGREDIENTS ~
POWER PAPER (One that already has a superpower trapped inside)
2 Tablespoons DRAGON SPIT (white vinegar)
OTHER SUPPLIES: Small container, Magic wand

~ Time: 5 minutes ~

1. Put 2 Tbsp of **Dragon Spit** into a small container.

2. Get your Power Paper. The writing should be a dark red, which means the superpower is trapped inside. Set the paper on a counter and turn it so that the top is facing East—in the direction the sun rises.

3. Now for the spell. Wave your wand over the paper 3 times in a swish and flick motion and say the words of the following spell: (Fill in the blank space with the name of the superpower)

*The power to _____ will soon be mine, for I have made this spell so fine.
With this power, I'll be just and true. Not like the dark wizard You Know Who.*

4. Now to absorb the power. Dip your thumb in the **Dragon Spit** and then press it to the paper right on top of the red thumbprint. Rock your thumb back and forth to make sure you completely cover the old thumbprint. Pull your thumb away. If the thumbprint magically disappeared, then you have absorbed the power into your body.

Experiment

Try dipping your entire hand into the magic ink and making a hand print on the paper. Once it is a dark red color, dip your hand in Dragon Spit, and press it against the red handprint. Hold your hand there and see if you can feel the magic soaking into your hand. The paper may feel like it is fizzing against your skin.

Transfiguration Potion

Make sure you speak clearly. My friend was doing this potion to impress Jessica Pillhimer she's the prettiest girl in our class and she loves dogs, but Max didn't speak clearly. Instead of turning into a dog, he turned into a hog. He was a big fat pig for 3 days!

This transfiguration potion has been used for hundreds of years. We even find evidence of its use in nursery rhymes where animals can speak or have other human traits. *Baa Baa Black Sheep*, *Hey Diddle Diddle*, and *Pussy Cat, Pussy Cat* are all nursery rhymes about people who have been transformed into animals. There is also evidence that *the little old woman who lived in a shoe* took some of her many children and used this potion to turn them into *the five little monkeys who jumped on the bed*.

What You Will See
Several different colored liquids stacked on top of each other.

What It Is Used For
This potion can be used to turn a person into an animal. It is also nice to look at while singing *Somewhere over the Rainbow*.

Ingredients

Toxic Fluid — 1–3 drops — Green food coloring	**Dragon Blood** — 3–6 drops — Red food coloring	**Troll Tears** — 1–3 drops — Blue food coloring
Tubeworm Pus — ½ cup — Dawn liquid dish soap	**Spider Venom** — ½ cup — Light-colored corn syrup	**Oil of the Salamander** — ½ cup — Cooking oil
Death Tincture — ½ cup — 70% or higher Rubbing alcohol	**Water** — ½ cup	

Other Supplies
- 1 CLEAR container (a tall glass or jar works well)
- 5 paper cups
- Pen or marker
- Turkey baster (a spoon can also be used)
- Stirring spoons
- Measuring cups
- Small objects such as: penny, pop tab, candy, plastic bead, small plastic lid, popcorn

Instructions

~ Time: 25 minutes ~

You will be making a 5-layer potion. Each layer will be a different color and will represent a different animal. The first 4 layers are Fox, Eagle, Dragon, and Lion. The last layer will represent an animal of your choice.

1. Select a clear container for your potion. Tall and narrow is best, and it should hold about 3 cups of liquid.

2. Set out 5 paper cups and use a pen or marker to label them: **Spider Venom**, **Tubeworm Pus**, **Water**, **Oil of Salamander**, and **Death Tincture**. Pour ½ cup of each of those liquids into their assigned cup.

3. Put 3 drops of **Dragon Blood** into the **Spider Venom** and stir to mix.

4. Look at the color of your **Tubeworm pus**. If it is white or clear then add 2 drops of **Troll Tears** and stir to mix. If it already has a color then don't add anything to it.

5. Add 1 drop **Dragon Blood** and 1 drop **Troll Tears** to the cup with the **Water**. Stir to mix so that you have a purple color.

6. Add 3 drops of **Toxic Fluid** to the cup with the **Death Tincture** in it. Stir to mix.

7. Carefully pour the **Spider Venom** into the tall, clear container. Pour slowly and try not to let it touch the sides of the container.

8. Now use a turkey baster to add the rest of the liquids in this order:

    Tubeworm Pus
    Water
    Oil of Salamander
    Death Tincture

9. Try to add the layers gently so they don't mix too much. If you don't have a turkey baster, another way to add the liquid is to gently pour it over the back of a spoon.

10. When you are done, get down and look at the container at eye level to see the layers clearly.

~ HOW TO TRANSFIGURE INTO AN ANIMAL ~

Take a small object and hold it tightly in one hand. Choose something that you do not have a personal attachment to such as a soda can tab, a plastic bead, a penny, a cherry tomato, a small plastic lid, a small piece of candy, a popcorn kernel, or a grape stem. Then say the following spell and drop the object into the transfiguration potion.

Colors of the rainbow in my potion bright,
tell me which animal I should be tonight.

THE LAYER IN WHICH THE OBJECT STOPS IS THE ANIMAL YOU WILL BE.

From the bottom up:
1) Bottom/first layer (red) FOX
2) Second layer (yellow, orange, blue, or green) CHICKEN, TURKEY, EAGLE, or DUCK
3) Third layer (purple) DRAGON
4) Forth layer (light yellow) LION
5) Fifth layer (green) ANIMAL OF YOUR CHOICE

Sometimes an object will float between two layers, but look closely and you will usually find that it is setting in one layer more than the other. If it is exactly between two layers then you may become a mix of both animals.

Now complete the transformation by repeating the following words, filling in the blank with the name of the animal:

The potion has spoken and I agree,
a _____ is what I shall be.

~ HOW TO DO THIS POTION ON SOMEONE ELSE ~

When turning someone else into an animal, give them the small object to hold and then wait for them to set it down. Do not touch the object yourself after they have held it. Use gloves or a paper towel to pick the object up. You may then repeat the following spell, inserting the person's name in the blank, and then drop the object into the potion.

Colors of the rainbow in my potion bright,
tell me which animal (person's name) should be tonight.

Look where the object lands in the potion to determine what animal they will become. Now finish with the following spell, inserting the name of the animal and the person's name in the last line:

The potion has spoken and I agree,
a (animal) is what (name) shall be.

Within a minute or two, the person should start acting like the animal. If your wizarding powers are strong, then the person will completely turn into an animal. If you are a beginning wizard and haven't developed your powers yet, the person will only take on a few characteristics of the animal.

— Bonus Activity —
Magical Stones

Ingredients

Other Supplies

- Microwave
- Microwave-safe bowl
- Strainer
- Paper towels
- Measuring spoon
- Measuring cup
- Cookie cutter (optional)
- Things to decorate with: markers, glitter, paint, etc.

| 1 ½ cups | **Unicorn Milk**
Milk–skim works best |

| 4 tsp | **Dragon Spit**
White vinegar |

Adult Wizard Info

~ WARNING ~
This potion involves using a microwave oven and should only be done with the help of a grown-up wizard.

Instructions

~ Time: 30 minutes ~

1. Pour 1½ cups of **Unicorn Milk** into a microwave-safe bowl, then add 4 teaspoons **Dragon Spit**.

2. Put the mixture in the microwave and cook on high for 1½ minutes.

3. Have an adult remove the bowl from the microwave and pour the mixture into a strainer.

4. The strainer will catch the glob—which you will use to make your magic stones. Let the glob cool then dump it onto a paper towel. Use more paper towels to press out any extra liquid.

5. Divide the glob into 3 pieces and shape them into stones. If you want a special shape, you can press the glob into a cookie cutter.

6. Let them dry for 2–3 days. Turn them over at least once during the dry time so that all sides get exposed to the air. Drying time will depend on how thick the stones are.

7. After a few days they should be hard—just like stone. Now you can use markers, glitter, or paint to decorate them how you want.

8. After they are decorated, you can give the stones magic by repeating the following spell:

I turned this liquid into stone, a gooey mess now hard as bone.
With this spell, I give it power. Now let it start this very hour.

~ NOTE ~
If you want to give the stones additional power then you must perform the spell under the light of a full moon, standing next to a large tree (preferably a sycamore tree), facing east, while wearing mismatched socks.

Potion Ingredients List

Name of ingredient	What it is	Where to buy it
Corpse Dust	Stump Remover (Potassium Nitrate)	Online (Amazon, eBay, etc) Home Depot and Lowe's both carry "Spectracide Stump Remover"
Crumbs of Alligator Skin	Yeast	Grocery stores in the baking isle
Curse Paper	Tea Bags	Grocery stores
Death Tincture	70% or higher Rubbing Alcohol	Any supermarket in the pharmacy section
Dragon Acid—Extra Strength	20 volume Hydrogen Peroxide Liquid	Beauty supply stores or hair salons
Dragon Blood	Red Food Coloring	Grocery stores in the baking isle
Dragon Spit	White Vinegar	Grocery stores
Dragon Stomach Acid	Hydrogen Peroxide 3%	Supermarkets in the pharmacy section
Fairy Wing Powder	Powdered Sugar	Grocery stores
Fire Fluid	Lemon Juice	Grocery stores
Fizzy Fizzing Fizzbee Tablet	Alka-Seltzer or Antacid tablets	Supermarkets or pharmacies
Flickerboom Tablet	Glucose Tablets or Dextrose Powder	Supermarkets or pharmacies
Graveyard Dust	Cream of Tartar	Grocery stores
Ground Dragon Egg	Sugar	Grocery stores
Ground Snake Fang	Sodium Hydroxide (Lye or "Caustic Soda")	Online (Amazon, eBay, etc.) Chemical suppliers. Some hardware stores as a drain opener. (Lowe's usually carries "Crystal Drain Opener" made by Roebic Laboatories, which is 100% Sodium Hydroxide.)
Octopus Powder	Borax	Supermarkets in the laundry detergent section
Ogre Sweat	Household Ammonia	Supermarkets
Oil of Salamander	Cooking Oil	Grocery stores
Orange-Thistle Tablets	Vitamin C Tablets	Supermarkets or pharmacies
Pixie Soap Bar	Bar of Ivory Soap	Supermarkets
Pond Slime	Clear Elmer's Glue	Supermarkets
Powdered Unicorn Horn	Baking Soda	Grocery stores
Scarab Beetle Powder	Boric Acid	Pharmacies (If they don't have it on hand, they can order it in for you)
Shifting Mists	Methylene Blue	Online (Amazon, eBay, etc.) Some aquarium stores that sell fish and fish supplies
Sneezewort Powder	Turmeric Powder	Grocery stores near the spices
Spell Paper	Paper Towels	Grocery stores/supermarket
Spider Venom	Corn Syrup (light colored)	Grocery stores
Spirit Dust	Corn Starch	Grocery stores
Supernatural Snot	Yellow Food Coloring	Grocery stores
Toxic Fluid	Green or Neon Green Food Coloring	Grocery stores
Troll Tears	Blue Food Coloring	Grocery stores
Tubeworm Pus	Liquid Dish Soap (Dawn works best)	Supermarkets
Unicorn Milk	Milk (2%, whole, or skim depending on which experiment)	Grocery stores
Vampire Venom	Liquid Iodine Solution	Supermarkets or pharmacies
Zombie Acid	Menthol Alcohol (Heet™ in the yellow bottle)	Supermarkets, gas stations, auto-supply stores